SPECIAL OPS

JOURNAL OF THE ELITE FORCES & SWAT UNITS

VOL.41

CONCORD
PUBLICATIONS COMPANY

10/F, B1, Kong Nam Industrial Building
603-609 Castle Peak Road
Tsuen Wan, New Territories
Hong Kong
www.concord-publications.com

We welcome authors who can help expand our range of books. If you would like to submit material, please feel free to contact us.

We are always on the look-out for new, unpublished photos for this series. If you have photos or slides or information you feel may be useful to future volumes, please send them to us for possible future publication. Full photo credits will be given upon publication.

ISBN 962-361-139-0
printed in Hong Kong

p3
Finns Wearing Fins
Finland's Erikoistoimintaosasto Navy Combat Divers
Juan Pablo Lasterra

p14
"There's Nobody Quite Like Us"
Gurkhas in Afghanistan
Carl Schulze

p32
Greek Police EKAM Counter-Terrorist Unit
Juan Pablo Lasterra

p44
Wielding an Iron Sword
NATO Reaction Force in Norway
Carl Schulze

Finns Wearing Fins

Finland's Erikoistoimintaosasto Navy Combat Divers

Juan Pablo Lasterra

In this picture we can see different uniforms, weapons and gear in use with the Erikoistoimintaosasto. These include equipment for maritime counterterrorism, combat diving and sniping. Even though they are a relatively small force, Finnish combat divers are second to none in their terrain and are prepared to operate in the coldest waters of the world.

"Special Ops Journal" visited a Scandinavian military special unit little known by the foreign public. The men who compose the Finnish Navy Special Operations Force have been arduously trained to complete their tasks in the most adverse environmental conditions.

If the waters surrounding any country's territory have a crucial commercial and military significance, then Finland must surely be one of the prime candidates. In Finland's case we can say that the control of its coasts has always been imperative to its survival. The country shares borders with Norway, Sweden, Russia and the Baltic states, a strategic area of permanent interest.

The Gulf of Finland Naval Command is based in Upinniemi, near Helsinki. It includes, amongst other units, the 5th Mine Warfare Squadron into which the Divers' Unit is integrated. It is divided into two different entities. The first is the Navy's Diving School (Merivoimat Sukeltajakursii), which was created in 1954 and charged with instructing Finnish Navy divers. This duty is carried out through two different courses.

- The Mine Countermeasure course (MCM) lasts 12 months and instructs students in the art of underwater mine-clearing.
- The Combat Divers course, as the name indicates, teaches divers to conduct special operations. These men compose the second group of the Divers' Unit - the Merivoimiexn Erikoistoimintaosasto, which translates as Special Operations Force of the Navy.

Progressive Training

The Finnish Navy (Merivoimat) began forming its combat divers' unit from the 1950s through to the 1970s, though it was not until 1993 that it was considered as an official unit by the Ministry of Defense. This is the reason why there is no public documentation about the subject. At the present time, their assignments are special reconnaissance, direct action (mainly carrying out raids, sabotage and maritime antiterrorism duties), and military support to other Finnish agencies. All these missions are executed within the framework of national defense or multinational operations. Let us remember that Finland is a European Union (EU) member not aligned with NATO, though it frequently participates in EU operations related to the resolution of crises in areas of conflict. Finland also actively collaborates with NATO and maintains bilateral cooperation with Russia.

The metal insignia of Finnish Navy Divers (top) and a freefall silver award worn on the jacket of a Finnish combat swimmer. The Erikoistoimintaosasto force did not have its own badge or insignia when we visited the unit, but the commander told us that some designs were being evaluated for adoption.

Two pairs of combat divers equipped with swimming gear and "ghillie" suits pose for the camera at their garrison at Upinniemi, near Helsinki.

As in nearly all Special Operations Forces, a pair is the smallest component of an operative unit. Note the waterproof radio system used by the operator on the left.

The Erikoistoimintaosasto responds directly to the Merivoimat Naval Headquarters and is composed of professional and conscript members, although a process focused on producing an entirely professional unit began in 2000. All divers are volunteers and every year the candidates must complete three days of selection before being admitted into the Combat Divers course.

The commander of the Erikoistoimintaosasto gave us some information about this selection process: "It's composed of several physical tests, among them a 15km forced march, 12 minutes of running, a kilometer of swimming with fins, a test in the diving tank, and a series of pushups and sit-ups. There's also an interview, a psychological test, and a medical examination. After that, selected candidates begin six months of training, considered as a qualification course, in which they must pass four levels of increasing difficulty before obtaining their Combat Diver insignia. The last one of these levels lasts for three uninterrupted days of swimming, diving, marching, and canoeing."

"After this basic training," continues the commander, "the divers specialize in one of the following skills: signals, weapons, engineer or medic. They learn to dive using Dräger LAR-V closed-circuit breathing systems, learn small-unit tactics, reconnaissance, raids, demolition, land and sea navigation, small-boat operations, close-quarters battle (CQB), hand-to-hand combat, and other special skills ."

Front view of a Finnish combat diver as he emerges from the cold waters of Upinniemi. Note the waterproof submersible communications equipment.

A Small Professional Force

Entering the "active unit", the main core of the Erikoistoimintaosasto, is reserved only for professionals. This small force completes another 12 months of additional training, including a parachute course in the specialist Finnish Army school in the vicinity of Utti.

The Erikoistoimintaosasto trains regularly with the Special Jaeger Battalion (Special Forces from the Finnish Army's Utti Jaeger Regiment), as well as Osasto Karhu (Bear Force - the Finnish Police special unit). It also maintains close ties with Swedish and Norwegian combat diving units. When necessary for training or missions, the Helicopter Battalion of the Utti Jaeger Regiment contributes helicopters.

We did not want to leave Upinniemi without asking the commander about how some of the coldest waters in the world affect Erikoistoimintaosasto in its training and operations. "The frozen climate has a great impact on preparations for training and in the use of equipment. The gear is evaluated to verify it can be used in temperatures below zero. But the main key to success, in my opinion, is our hard and continuous training, because our equipment is not very different to that used in similar units worldwide. So we put emphasis on the individual ability of soldiers to stay ready for combat in this hostile environment, especially in winter."

Erikoistoimintaosasto operator looks for signs of a hostile presence during a demonstration of coastal reconnaissance. He is armed with the locally produced Sako RK95 7.62 x 39 assault rifle.

Polar underwater "dry suits" are impermeable coveralls that allow the diver to wear his uniform and tactical vest underwater without them becoming wet. The breathing apparatus is the Dräger LAR-V, a closed-circuit system that does not emit bubbles underwater, thus ensuring a swimmer's movements remain clandestine.

A Zodiac 470 inflatable boat provides momentary cover for these two combat swimmers during an exhibition showcasing "wet" infiltrations.

Erikoistoimintaosasto is a professionally trained and equipped unit. Finland is situated in an area of critical strategic interest to Europe. The operations performed by the force and many details about Erikoistoimintaosasto are kept in extreme secrecy by the Finnish Ministry of Defense.

Weapons and Equipment

The weapons in service with the unit are FN HP DA pistols in 9mm Parabellum (with the local designation of P-80); H&K MP5 Navy standard and silenced sub-machineguns in the same caliber (some samples have been equipped with diurnal Aimpoint sights or Patria VV2000 night sights, tactical lights and lasers); Sako RK95 assault rifles in 7.62 x 39 caliber (with the same accessories as the MP5); SAKO TRG-42 sniper rifles in .338 Lapua Magnum caliber fitted with a Carl Zeiss scope; PKM machine guns in 7.62 x 39 caliber; and single-shot H&K GZP-1 40mm grenade launchers.

The underwater "dry suits" are made by Polar. These impermeable coveralls allow the diver to wear their uniform and tactical vest underwater without them becoming wet. The breathing apparatus in service is the Dräger LAR-V, which uses a closed-circuit system that does not emit bubbles underwater. This maintains the secrecy of a swimmer's movements. The Erikoistoimintaosasto also uses standard Finnish Armed Forces camouflage uniforms, Sacci, Savotta and Karrimor rucksacks, Karrimor Chameleon "ghillie" suits, Burgmann ballistic vests, Savotta tactical gear, standard M/05 Kevlar helmets, Simrad and ITL Mini N/SEAS night vision goggles and thermal viewfinders, plus other miscellaneous items. Some of the communication radios in service are completely submersible.

For transportation, the unit uses Avon and Boomeranger rigid inflatable boats (RIB); Zodiac 470 inflatable boats; Alutech Watercat M8 vessels (also known as G-Boat) that have an aluminum hull, jet propulsion and an eight-passenger capacity; and kayaks. Some of the aforementioned boats mount an Utjos NSV 12.7mm Russian heavy machinegun.

The Finnish Navy

The 39,000 men of the Merivoimat are organized into eight main components:

1. Naval Headquarters, located in Helsinki
2. Archipelago Sea Naval Command. Based in Turku, it includes the following units:
   - 6th Missile Squadron
   - 4th Mine Warfare Squadron
   - Turku Coastal Battalion
   - Archipelago Sea Base Battalion
   - Navy Band
3. Gulf of Finland Naval Command. Located in Upinniemi, it includes:
   - 7th Missile Squadron
   - 5th Mine Warfare Squadron
   - Naval Training Center
   - Hanko Coastal Battalion
   - Gulf of Finland Base Battalion
4. Kotka Coastal District. Located in the region of the same name, it is made up of two coastal forts in Kirkonmaa and Rankki
5. Uusimaa Brigade. With a base in Dragsvik, it is composed of:
   - Vaasa Coastal Jaeger Battalion
   - Raseborg Coastal Artillery Battalion
6. Naval Academy, in Helsinki
7. Naval Research Center, in Espoo
8. Naval Material Command, in Turku.

In this image we can see different details of individual diving equipment used by Finnish combat swimmers. Note the nonmagnetic knife, the submersible communications gear, the tactical vest, and an underwater compass board on the ground.

Two combat divers remove their gear after finishing a demonstration in Upinniemi, the location of the Gulf of Finland Naval Command's base.

A team of combat divers approaches the coast of Upinniemi at high speed using one of the Zodiac 470 boats they have in service.

An Erikoistoimintaosasto team approaches the coast of Upinniemi using a camouflaged Zodiac 470 boat.

Pointing their Sako RK95 assault rifles towards possible threats, an Erikoistoimintaosasto team prepares to disembark from their Zodiac 470 boat. The Sako RK95 7.62 x 39 assault rifle is a locally made copy of the famous Kalashnikov. As with all Finnish weapons, it is a well-built rifle that can operate without malfunctions in a wide range of climates, especially in the cold waters and frozen northern areas of the country.

Wearing tactical vests over their Polar dry suits, a reconnaissance group from the Finnish Navy Special Operations Force approaches its objective during a demonstration put on especially for the benefit of this article. The Polar dry suit also permits operators to wear their tactical vest underneath the system.

A 4-man team prepares for a diving session in cold and murky Finnish waters. (Finnish Merivoimat)

The Valmet RK62 assault rifle was in service when this picture was taken. Nowadays, the unit only uses the more modern RK95 7.62 x 39 weapon in all its exercises. This weapon is in service with all branches of the Finnish armed forces. (Finnish Merivoimat)

After submerging, these combat swimmers dive towards their objective with the aid of a submersible compass board. (Finnish Merivoimat)

For fast transportation, Erikoistoimintaosasto combat divers use Avon and Boomeranger rigid inflatable boats (RIB), as well as other advanced craft. (Finnish Merivoimat)

The zone around Upinniemi, where the Gulf of Finland Naval Command and the combat divers are based, is excellent for practicing deployments with all the vessels and aircraft that the Erikoistoimintaosasto has at its disposal. (Finnish Merivoimat)

The Sako RK95 assault rifle is gas operated with a rotating block action. It chambers the 7.62 x 39 Russian cartridge. The trigger guard is enlarged to enable shooting using diving or winter gloves.

A pair of Erikoistoimintaosasto operators during a maritime counterterrorism exhibition inside a Finnish Navy vessel.

Erikoistoimintaosasto operators are trained and equipped to carry out different missions that range from classic coastal reconnaissance tasks to raiding enemy ports or maritime counterterrorism.

Two Erikoistoimintaosasto operators storm a room as they search for possible terrorists during a maritime hostage rescue exhibition.

A pair of operators "clean" a vessel deck. Their kit is a mix of Finnish Navy standard gear with foreign elements, as illustrated by their German H&K MP5SD-Navy sub-machine guns.

An Erikoistoimintaosasto operator waits his turn during a counterterrorism drill on a Finnish Navy vessel.

A "maritime" variant of the SD6 model, the German H&K MP5SD-Navy silenced sub-machine gun in service with the Erikoistoimintaosasto was specifically conceived for use in humid environments.

The FN HP DA pistol in 9mm Parabellum is in use with all Finnish armed forces – it has the local designation of P-80. It is possible that Erikoistoimintaosasto will replace this venerable sidearm with a more modern model in the future, following the example of Finnish Army Special Forces.

In this view we can see the position of the communications system and radio antenna during operations in close environments such as buildings or vessels.

Using a thermal viewfinder, the spotter of an Erikoistoimintaosasto sniper team looks carefully for the best target for his comrade. The quality of each sniper shot depends in great measure on the data provided by the spotter. The spotter is charged, amongst other tasks, with measuring the distance to the target and the velocity of the wind.

A pair of snipers perfectly camouflaged in the forest near Upinniemi. Note the use of natural concealment, as well as Karrimor Chameleon "ghillie" suits. The shooter is aiming his Sako TRG-42 rifle in 338 Lapua Magnum.

The spotter of a sniper team points his Sako RK95 assault rifle that is partially camouflaged in green tape. A number of diurnal Aimpoint sights, Patria VV2000 night sights, tactical lights and lasers are available in the unit's armory for these weapons.

Portrait of an Erikoistoimintaosasto sniper. He is dressed in a Karrimor Chameleon "ghillie" suit made in the United Kingdom. This outfit not only helps the user to "melt" into his surroundings, but also reduces his thermal signature. Note how this sniper has also camouflaged the lines of the Sako TRG-42 sniper rifle with locally applied foliage and camouflage tape.

A pair of Erikoistoimintaosasto snipers advancing through the forest. The spotter moves first, armed with a Sako RK 95 assault rifle. Both operators are dressed in Karrimor Chameleon "ghillie" suits, which are manufactured in various camouflage tones (forest, desert, snow etc.) for different types of terrain.

Finnish combat divers can perform reconnaissance and raid missions not only in the sea, lakes or rivers, but also on dry land. The most recent conflicts such as those in Kosovo, Afghanistan and Iraq, have seen combat swimmers (for example, U.S. Navy Seals, British SBS, Italian Comsubin and French Hubert commandos) executing their tasks hundreds of kilometers away from any "wet" zone. Only one of the three aforementioned national territories actually even has a coastline! (Finnish Merivoimat)

"There's Nobody Quite Like Us"
Gurkhas in Afghanistan

Carl Schulze

Mazar-i Sharif: soldiers of the QRF Company based at Himalaya Lines provide security while ISAF flight operations are conducted at the local airport. Note the C-160 Transall aircraft of the German Air Force that has just landed. Like the KPC in Kabul, the QRF Company in Mazar-i Sharif uses the Land Rover XD110 WMIK as its main patrol vehicle. The pictured vehicle has a 7.62mm GPMG on its weapon ring mount.

Since they first enlisted with the British Army in 1815, Gurkhas have forged for themselves a reputation as elite troops. During almost 200 years of service for the British Crown, these soldiers from Nepal have proved themselves to be skilful, reliable and fearless warriors. In countless conflicts such as WWI (1914-18), WWII (1939-45) and the Falklands War (1982), the Gurkhas were used as shock troops. Therefore, it is no wonder that the 2nd Battalion, The Royal Gurkha Rifles, deployed to Afghanistan as part of British Forces Afghanistan. In Afghanistan the Gurkhas again proved their value. With professionalism and enthusiasm, the Gurkhas adapted themselves well to their mission as a peace support force. The fearless fighters from the Himalayas, famed for their kukri knives, showed in Kabul and the area around Mazar-i Sharif how well they are suited for peace support operations.

Operation Herrick

The 2nd Battalion, Royal Gurkha Rifles (2 RGR), is currently based in Brunei. The unit is the only one within the British Army permanently trained in jungle warfare. In addition, it is the only British battalion-sized combat formation in Southeast Asia available at short notice for deployment during a crisis or to protect British interests in the region. Due to this fact, the battalion is kept at a constant state of readiness and

The Gurkhas are skilful warriors. In their inventory they combine modern weapons such as the 5.56mm MINIMI light machine gun with ancient weapons such as the feared kukri knife. Here a soldier of 2 RGR can be seen posing for the camera with his kukri drawn ready for action.

Like their ancestors in the Afghan Wars, the Gurkhas still carry the kukri knife as a sidearm. Here an officer of the QRF Company at Mazar-i Sharif can be seen posing for the camera with his kukri. Note its size. It has to be mentioned that most soldiers of 2 RGR in Afghanistan carry kukris twice as long as those normally seen in peacetime.

maintains a high level of training. After several months of preparatory training in Brunei, the unit deployed to Afghanistan for a six-month tour in early April 2005. For the battalion it was their second deployment to Afghanistan as two of its companies had already deployed to Kabul between September 2003 and April 2004. It was during this tour that Colour Sergeant Kajiman Limbu rescued an injured U.S. lieutenant colonel while under enemy fire. Kajiman Limbu was awarded the Military Cross for his gallant action. As part of Operation Herrick, 2 RGR became the Afghanistan Roulemont Infantry Battalion and was placed under the NATO-led International Security Assistance Force (ISAF). The mission of ISAF, and therefore of 2 RGR, is as follows:

- Reassure and help the people of Afghanistan to return to normality after years of war.
- Support the creation of new Afghan security forces including police and the Afghan National Army (ANA).
- Support the government of Afghanistan in creating a safe and secure environment in order to reinstate the rule of law.
- Enable Afghanistan to be a prosperous, democratic and successful country free from terrorism and conflict.
- Deter terrorism.

Within this mission, the tasks of 2 RGR are enormous and the battalion is split into Operation Herrick (North) and Operation Herrick (South). Under Operation Herrick (North), 2 RGR runs a Forward Support Base (FSB) near Mazar-i Sharif that supports the British-led Provincial Reconstruction Team (PRT) in the city (also manned by soldiers of 2 RGR). It also maintains a company-sized Quick Reaction Force (QRF) that can be deployed anywhere in the whole of northern Afghanistan, plus it fields several six-man Military Observation Teams (MOT) that operate in the remote mountainous areas of northern Afghanistan. Except for a platoon based at the PRT in Mimahna and the MOTs based in safe houses all over the Area of Operations (AOR), the bulk of the Gurkhas under Operation Herrick (North) are based at Camp Himalaya Lines near the airport just outside Mazar-i Sharif.

In Operation Herrick (South), 2 RGR supports HQ British Forces (BRITFOR) Afghanistan, maintains the UK National Support Element (NSE), runs the Afghan National Army Training Team (ANATT), and provides the Kabul Patrol Company (KPC). The KPC is under the command of the Kabul Multinational Brigade of ISAF and is responsible for security in the city's Police Districts 8, 9 and 16. All Gurkha assets in Kabul are based at Camp Souter. The camp is situated north of the Jalalabad road close to Kabul International Airport. Without question, the 500 Gurkhas and their British officers have a busy time with all these different commitments.

Here it needs to be added that the situation in Afghanistan is far from stable and safe. In order to illustrate the security situation we will mention some of the incidents that occurred while the author visited Kabul in late May 2005. On 17 May, a female Italian aid worker of Care International was kidnapped from her home, an Improvised Explosive Device (IED) was found on the road leading to Camp Black Horse (the Canadian base in Kabul), an IED team member was killed during mine-clearing operations at Kabul International Airport, and a hospital was narrowly missed by two 107mm rockets fired during an attack. Reports state that some 16 children are kidnapped in Kabul in an average week. In the early morning hours of 30 May, a 107mm artillery rocket was fired at the ISAF HQ in Kabul in the city center.

In Afghanistan the Gurkhas field weapons that have recently entered British armories. Here a soldier of the Mazar-i Sharif-based QRF can be seen on guard duty. He is armed with a 5.56mm SA80A2 under which a 40mm AG36 grenade launcher is mounted. The AG36 is made by Heckler & Koch, has a weight of 1.5kg, is fitted with a ladder sight, and has a combat range of 350m. The design of the AG36 with the break-action barrel allows it to fire all sorts of 40mm ammunition, even long cartridges that could not be used with the U.S.-made M203 grenade launcher.

A KPC patrol has stopped at an Afghan police checkpoint in southern Kabul so that Gurkha patrol leaders can exchange information with the policemen. The mission of the patrol is to gather information about a suspected plane crash that later turned out to be a false alarm. However, an accident involving an aircraft landing or taking off is not farfetched as Kabul is surrounded by high mountains. In early 2005, a passenger aircraft crashed in the mountains southeast of Kabul.

The rocket hit an accommodation building in the ISAF compound. Luckily, nobody was injured. During the day, a bicycle-based IED was detonated at the entrance to Camp Phoenix (a U.S. camp situated on Route Violet leading from Kabul to Jalalabad). This explosion injured several Afghan civilians.

Kabul Patrol Company

The patrol moves slowly down the muddy road in Police District 9 of Kabul. Alongside the road runs a small ditch filled with sewage. The temperature is 29°C, and the smell of the sewage far outweighs the smell of spices sold at one of the countless market stalls lining the road. The street is crowded with females wearing blue-colored *burkhas* as they make their daily purchases, and men negotiating prices at market stalls. A little further down the road, a local is spreading water from the sewage ditch onto the road in order to prevent the dry ground from becoming a dust cloud whenever vehicles pass by.

This Land Rover XD 110 WMIK (Weapons Mounted Installation Kit) belongs to the QRF of 2 RGR. The vehicles were seen during a mission where the QRF was called to the site of a mine explosion. It turned out that a cow had walked into a marked minefield near Kabul Airport and set off a mine. The cow was killed but nobody else was injured!

Members of the 2nd Battalion, The Royal Gurkha Rifles, are running a mobile vehicle checkpoint in the city of Kabul. While the Gurkhas provide security for the mobile VCP, Afghan policemen conduct vehicle searches. However, the patrol leader and his interpreter are assisting the policemen by providing useful and much-needed advice.

During a live-fire exercise, a Gurkha instructor of the ANATT acts as a fire team leader in order to assist the Afghan section commander. Here the Gurkha directs covering fire provided by his fire team while the other team can be seen advancing in the background.

The Gurkhas move through the crowd, carefully keeping each other in view. The patrol leader is accompanied by an Afghan police officer, a common practice as most patrols of the Kabul Patrol Company (KPC) are joint. An interpreter is also part of the patrol and he closely follows the patrol leader. Then the patrol stops at a small shop selling Bollywood movie DVDs and music CDs. Loud bhangra music emanates from a small stereo placed on the shopkeeper's desk. The patrol leader quickly starts a conversation with the shopkeeper - the language used is Urdu. Urdu is similar to Hindi, a language all Gurkhas speak fluently due to the fact that Nepal has a border with India and Hindi is the language of Bollywood films, which are as popular in Nepal as they are in Afghanistan. At first the shopkeeper is surprised that the foreign soldier can talk with him, but soon the ice is broken and the conversation turns from films and music to daily life, local political opinions, their thinking on ISAF, and the shopkeeper's feelings about security.

It is during this type of reconnaissance that the Gurkhas gather vital information from time to time. In one case, information about a group of thieves breaking into empty houses was gained. In another, rumors about a planned demonstration were collected. While the commander of the patrol is speaking with the shopkeeper, some of his patrol members start similar conversations with youngsters following their patrol around the streets of Kabul. Often the kids come up with information that backs up the facts given by adults. After a couple of minutes, the patrol sets off on its route once more. Again the Gurkhas have made friends and bolstered the confidence of Kabul inhabitants towards ISAF troops.

The KPC is based at Camp Souter close to Kabul International Airport. It is formed from A Company, 2 RGR, which is reinforced by a rifle platoon of B Company and a mixed Support Weapons Platoon with mortars, Milan antitank guided weapons and GPMGs from the battalion's Fire Support Company. These provide the KPC commander, Major Fraser Lawrence, with a formidable force of some 150 men in five operational platoons. "Rather then operating in sections, we operate in multiples in Kabul," Major Lawrence explains. "Each platoon breaks down into two 13-soldier strong multiples. The multiple is also the standard patrol asset and we have at least one multiple on the ground at any one time. Usually our patrolling consists of personnel mounted on Land Rovers, foot patrolling, and conducting temporary vehicle checkpoints." The company commander continues: "In addition, my soldiers sometimes get special tasks - for example, gathering information about the population in a certain area or finding out more details about a certain building."

Often the Gurkhas going out on patrol are supported by dog handlers of BRITFOR Afghanistan. Sometimes joint patrols are conducted. An example of this is Gurkhas and Norwegian or German soldiers patrolling

A member of 2 RGR's ANATT instructs Afghan soldiers during an NCO course held at the KMTC. The aim of the ANATT is to train a new NCO corps that in the future will provide the backbone of the new Afghan Army. But the Gurkhas face several difficulties such as a lack of initiative among the Afghan soldiers. Another problem is the motivation of the Afghan troops, which often drops due to irregular pay, a lack of ammunition for live firing, or the fact that weapons and equipment are in a poor state.

A Gurkha instructor of the ANATT conducts safety checks prior to a mortar live-fire exercise of an ANA NCO course. The mortar is an 82mm M37 of Russian origin. In general it could be said that the equipment of the Afghan Army is in a poor state. For example, the traversing gear was not functioning on this particular weapon, though this does not prevent the Afghan soldiers from firing it. Stunning is the fact that even with this broken mortar, they managed to hit the target!

The MOTs operate in remote areas of northern Afghanistan. Mobility for the teams is provided by civilian cross-country vehicles. Usually a MOT consists of the MOT commander and his deputy, two drivers for the unit's cross-country vehicles, a signaler, a medic and an interpreter. The MOT commander is usually a young officer or a senior NCO, while the MOT mission requires team leaders who show a high degree of initiative, flexibility and responsibility.

together along the boundaries of their AORs. "My platoons rotate every couple of days through the different tasks - two platoons conduct patrols while another is providing camp security," explains the major as he elaborates on the routine of his Gurkhas. "The fourth platoon is on stand by as QRF for the Kabul Multinational Brigade (KMNB) under which we are subordinated, while the fifth platoon does maintenance work as well as conducting mission-orientated training." The KPC's QRF is usually called out several times a day. Suspected IED sites often have to be sealed but in most cases the suspected IED turns out to be a mistaken empty gas bottle or a discarded storage box. However, the IED threat in Kabul is becoming more and more serious and the Gurkhas take every alarm as an opportunity to hone their skills. "After the Koran scandal on 12-14 May we had some large demonstrations in Kabul. They were well organized and peaceful, but we kept the QRF in anti-riot gear in reserve," recalls the company commander. "The QRF's prime mission is to deploy when the lives of members of the Coalition Force or embassy members are in danger. In other cases it is the job of the local police and the ANA to deal with the problem, and we only become involved if the local security forces cannot deal with the problem on their own."

Also based at Camp Souter are 2 RGR's combat service support elements. These include the Motor Transport Platoon, the Quartermaster Platoon, and so on. Together with assets of BRITFOR Afghanistan, these units form the British NSE responsible for all logistic aspects of Operation Herrick.

CIMIC – Civil Military Cooperation

CIMIC operations are basically of a humanitarian type but they also have a military aim. They are designed for the local population to forge trust in the military, and therefore lead to better cooperation between civilians and soldiers. Furthermore, certain CIMIC activities have the aim of enhancing force protection of operational troops. The CIMIC cell of 2 RGR is responsible for CIMIC operations in Police Districts 8, 9 and 16, the same districts patrolled by the KPC. In addition, the Gurkha CIMIC team works with the KMNB CIMIC cell, which runs projects all over the brigade's AOR. "We run several projects simultaneously. Our main focus is put on education, public health and civil infrastructure," explains Captain Danial Lama, 2 RGR's former Motor Transport Officer who today runs the CIMIC team. "Several of these projects are conducted with Non-Governmental Organizations (NGOs) and the UN. We organized reconstruction of the kitchen and kindergarten of Maiwand Hospital, and supported reconstruction of Khair Khana Hospital," explains the captain.

"On the educational side we supported several schools with their curriculum and delivered basic school equipment," continues Captain Lama. "Another project we supported was situated in the cultural and heritage area; we helped the Afghan National Museum to improve security and administration issues. This included construction of a new guard and administrative building, erecting a new perimeter wall, and the installation of security lights. Projects with a civilian and security background were the construction of 200m of road in Alu Khel village, and refurbishing the Pulae Parcha Bridge. The bridge especially provided our forces with an

With his troops scattered all over northern Afghanistan and Kabul, the battalion commander of the 2nd Battalion, The Royal Gurkha Rifles, is constantly on the move visiting his troops. In addition to exercising command, LtCol Rigden makes countless visits to his soldiers in Mazar-i Sharif, Kabul and safe houses in the Hindu Kush mountains. The visits are aimed at boosting the morale of his men and receiving first-hand accounts about the situation on the ground. Here the battalion commander can be seen visiting the ANATT in Kabul.

Around their base at Camp Himalaya Lines near Mazar-i Sharif, the Gurkhas conduct mounted and dismounted patrols. Here such a patrol can be seen in a dry riverbed south of the base. Weapons used by the Gurkhas include the 5.56mm SA80A2 assault rifle, 5.56mm LSW, and 5.56mm MINIMI machine gun.

From the top of a Land Rover Defender XD 110, a Gurkha armed with a 5.56mm MINIMI machine gun provides cover for his comrades. The soldier belongs to the QRF of the KPC and his comrades have just established a security perimeter around a possible mine incident location.

alternative road for crossing the Kabul River. In a joint Anglo-American venture we donated food and drinks to people in need, as well as clothing, medicine, toys, blankets and shoes. We also organized construction of a dam to prevent the Kabul River from flooding parts of Kabul inhabited by poor people." Even with all these successes, the CIMIC team does not tire of starting new projects. During the author's visit to Kabul, some nine new projects were already in the queue or waiting for funding. Future aspirations included a plan to deliver firewood to hospitals and kindergartens, sandbags for the Ministry of Rural Development, as well as the delivery of tools and equipment to the mayor of Kabul.

Training the new Afghan Army – Gurkha ANATT

"Down, down!" yells the 2 RGR corporal. The Afghan soldiers drop to the ground and raise their 7.62mm AK-47 rifles in order to provide covering fire for the other fire team. The air is filled with noise when the Afghan soldiers start opening fire at targets some 35m away. At the same time, the Afghan section commander rouses the other half-section and advances in the direction of the target.

It is the end of May 2005, and the Afghan soldiers are practicing fire and movement drills with live ammunition on the Afghan Army ranges just to the east of Kabul. The troops belong to one of the four Junior Non-Commissioned Officer (JNCO) courses held by the Kabul Multinational Training Centre (KMTC). The KMTC is supported by the Training Advisory Group, which consists of British, U.S., French and New Zealand instructors. One of the missions of ISAF and U.S. troops operating in Afghanistan under Operation Enduring Freedom (OEF) is to support the reconstruction of a capable new ANA. The ANA in the future should be capable of maintaining a stable and secure environment in Afghanistan, and of protecting the country against aggression from foreign forces.

In order to train the ANA for this mission, Task Force Phoenix was formed. Based at Camp Phoenix, the U.S.-led TF Phoenix consists of the Training Advisory Group, Embedded Training Teams (ETT), and Mobile Training Teams. The Training Advisory Group works at Camp Alamo and supports the activities of the KMTC. The ETT, provided by Germany, Canada and the U.S., supports the training of ANA battalions that have already reached operational status. For example, the solitary ANA tank battalion is supported by a German ETT. The ETTs stay with their assigned Kandak for a period of at least 24 months. After this the Kandak is put through a validation test, and the ETT is removed if the battalion passes the test. If the Kandak fails, the ETT stays with the unit. The Mobile Training Teams provided by Romania, the U.S. and Mongolia assist ANA units in specific specialized training. The MTT only supports a unit for a short period of time. In the command structure, TF Phoenix is subordinated under the Office of Military Cooperation (Afghanistan) (OMC(A)), which supervises as well as monitors activities of the Afghan Ministry of Defense. The OMC (A) is also involved in the training of the Afghan National Police and the Disarmament, Demobilization and Reintegration Program (DDR).

A Gurkha Saxon commander can be seen here manning the vehicle's 7.62mm GPMG. The picture was taken at Kabul International Airport (KIA). The Gurkhas use the Saxon wheeled APC to transport personnel between KIA and Camp Souter.

Training Advisory Group and UK ANATT

The Training Advisory Group is commanded by Colonel Jones of the U.S. Army. The colonel and his instructor teams support the KMTC. The KMTC is the major training establishment of the ANA. Basic, NCO, officer and staff officer training is conducted at the KMTC. Due to the fact that the ANA is under construction, basic training is provided for battalion-sized formations raised from scratch. The KMTC has access to a large training area to the east of Kabul that features 13 live-fire ranges for weapons like small arms, mortars and tank guns.

The Training Advisory Group is divided into different sub-training groups called ANATT (Afghan National Army Training Team). A U.S. ANATT is responsible for assisting Afghan instructors in conducting six-week basic training for fresh recruits, and the following six-week advanced individual training. A French-led ANATT supports the 11-week officer training courses, while the staff officer training is conducted by a multinational ANATT. The UK ANATT is responsible for NCO training. At the time of writing, personnel of the UK ANATT were provided by the 2nd Battalion, The Royal Gurkha Rifles. For the task, the battalion's Support Company attached its soldiers to other companies and formed a staff of 19 selected NCOs and officers to act as the UK ANATT. Led by the company commander of the Support Company, Major Thomson, the UK ANATT also consists of two attached U.S. and two New Zealand instructors. The UK ANATT assists in the following KMTC courses:

- Junior NCO course. Usually four courses are run concurrently at the KMTC. Each course is run by 13 Afghan instructors monitored by one Gurkha instructor. The course covers platoon tactics in offensive and defensive battles, ambush training, patrol procedures, reconnaissance activities and fighting in built-up areas (FIBUA). The six-week course finishes with a four-day final exercise.
- Combat Leader course. The five-week course focuses on tasks of the platoon sergeant and involves similar training elements to the JNCO course. Three courses are usually run at the same time. Each course is run by five Afghan instructors monitored by one Gurkha instructor.
- Senior Sergeant course. The course focuses on the tasks of a warrant officer. Currently the course is three weeks long, but in the future it will be extended to four weeks. Only one course is run by the KMTC at any one time. The course is run by seven Afghan instructors and is monitored by one New Zealand instructor.

Gurkha patrols regularly conduct observation stops at the Mausoleum of Nadir Shah, which is situated on a hilltop in Police District 8. The ruins of the mausoleum can be seen in the background. From here the Gurkhas can watch over nearly their whole area of operations.

But the most important task of the UK ANATT is not monitoring the courses but "training the trainer" (T3). In an eleven-week period, Afghan instructors run JNCO and Combat Leader courses under the watchful eyes of the Gurkhas. The aim of T3 is to train a new generation of instructors for the KMTC that have a high level of knowledge. This is necessary due to the fact that most of the current instructors are used to an old-fashioned

"Afghan" style of leadership that is not compatible with the modern battlefield. "Only 30 out of 100 potential future instructors reach the required standards," explains Major Thomson. "Potential instructors are selected by the Afghan MOD, and when they arrive here we put them through a two-day test. Normally only 40 are left after this."

Selection of NCOs

In addition to the T3 program, the Gurkhas also play an active role in the selection of future NCOs of the ANA. During the last week of basic training of a Kandak, the Gurkha instructors select some 155 soldiers with the potential of becoming NCOs. Key requirements for future NCOs are physical fitness, weapon skills, leadership capability, and most important of all, literacy. Here it must be stated that only 35% of Afghan people can read and write. Only 91 participants were literate in one of the last JNCO courses, but due to a shortage, illiterate participants were also accepted. The ANATT organized a basic writing and reading course for these soldiers. The selected NCO candidates are put through the JNCO course, and after successfully finishing it they will be assigned to a new Kandak that has just arrived in Kabul for basic training. The new JNCOs will now run basic and advanced individual training (AIT) of the new Kandak under the watchful eyes of Gurkha instructors. At the end of AIT, the new battalion will run an evaluation exercise. The results of this exercise are then passed to the ETT, which from then on monitors and assists the battalion's training.

Not an easy task

Major Thomson outlined some of the problems the Gurkhas encounter while monitoring ANA training. "We face a lot of challenges - some are related to a certain lack of equipment, others have to do with the mentality and habits of the Afghan population. For example, maintenance and keeping weapons clean is an uncommon practice in this country. Additionally, weapons often vary technically due to different manufacturers. Ammunition is another story - in some cases we get collected stuff from the DDR program, and often this ammunition is damaged or unserviceable." Major Thomson continues: "Corruption is also widespread in Afghanistan and we often have to check carefully that this is not affecting the training or promotion process. On the other hand,

Soldiers of the Mazar-i Sharif Gurkha element are seen here providing security while ISAF flight operations are conducted at the local airport. Like the KPC in Kabul, the QRF Company in Mazar-i Sharif uses the Land Rover XD110 WMIK as its main patrol vehicle. The pictured vehicle has a 7.62mm GPMG attached to its weapon ring mount. The soldier in the foreground is armed with a 5.56mm SA80A2 assault rifle.

motivation sometimes drops. Often the reason for this is that the government does not pay the troops on time. Another problem is that we send out a requirement for what kit the trainees should have when they arrive, but for various reasons this is seldom respected. We also face problems in terms of transport, as the number of available trucks is limited. Training sessions have to be adjusted to religious customs in Afghanistan and often the command and control system within the ANA is a challenge for us."

Major Thomson describes some of the other difficulties: "In addition, sometimes within the KMTC complex we don't have enough classrooms to conduct theory lessons. At the same time the hygiene standards are poor - at the dining facility, for example. This sometimes results in illnesses quickly spreading. All this means that the work of the Gurkhas is not easy, but the soldiers from Nepal have a major advantage that their comrades from other nations within the Training Advisory Team do not have - most Gurkhas speak Hindi or Urdu, and both languages can be understood by most Afghans. They therefore have an easy means of communicating with their trainees. For some of the other problems, the ANATT has found easy

A Gurkha instructor of the ANATT listens to an Afghan NCO briefing his fire team for a fire and maneuver exercise during an ANA NCO course. The fact that most Gurkhas speak Hindi and Urdu makes it easier for them to breach the language barrier between Afghan soldiers and their British instructors.

solutions. If soldier's weapons are found to be rusty, the NCO is punished with the soldiers. In order to prevent soldiers from leaving courses without leave, the ANATT organized that pay is only given a day before course graduation day. But even with these successful measures introduced by the Gurkhas, it will still be a long time before they break the old-style tribal thinking and form a new Afghan National Army."

This leads us to the matter of what happened to former Mujahideen. "The Mujahideen don't know the modern art of warfare. However, they have important knowledge about local tactics as well as knowing how to fight and survive in the extremely difficult Afghan environment," explains Major Thomson. "Therefore, they can become a JNCO without attending the course, although they have to prove their skill and we certainly won't allow them to also skip the Combat Leader course."

By the end of May 2005, personnel of 35 Kandaks had passed out of the KMTC. The units consisted of 650-800 troops each. The range of units included infantry, combat support, combat service support, armor, and Special Forces units. Current plans see the last of the 79 planned Kandaks finishing their training in 2007. The Kandaks are recruited regionally and mirror the ethnic mix of the regions they come from. Usually Kandaks include Tajiks as well as Pashtuns and soldiers of other ethnicities. When working together, the soldiers operate using Dari as the official ANA language. The ethnic groups usually stick to themselves off-duty due to different cultural customs and language problems. This fact clearly shows that the Gurkhas and their successors have a long road ahead in building a professional, democratic ANA capable of maintaining Afghanistan's security.

PRT, QRF and FSB in Mazar-i Sharif

Under Operation Herrick (North), the Gurkhas run the British PRT in Mazar-i Sharif. The mission of the PRT is to assist the Afghan government in extending its authority, and through a military presence, enabling security sector reform and reconstruction efforts in order to facilitate the development of a stable and secure environment in the AOR. To achieve this mission, the PRT runs countless activities ranging from organizing equipment for the local police to maintaining the Disarmament, Demobilization and Reintegration (DDR) process of former Northern Alliance forces.

While the PRT itself is based in the center of Mazar-i Sharif, the main base of 2 RGR is situated several kilometers outside the city at the airport. Called Himalaya Lines, this tidy camp features countless large tents that accommodate the Battalion HQ of 2 RGR, the Forward Support Base and the QRF. The FSB is maintained by elements of the battalion's HQ Company, as well as by the command group of C Company. Basically, the FSB supports the PRT and the MOTs through all possible means so they

While his patrol leader is having a chat with some locals during a patrol in Kabul, a soldier of 2 RGR keeps a watchful eye on the street. He is armed with a 5.56mm MINIMI light machine gun. Each infantry section of the 2nd Battalion, The Royal Gurkha Rifles, fields two MINIMIs.

Gurkhas of the Mazar-i Sharif-based QRF Company train on basic house entry skills in Camp Himalaya Lines. The first soldier is armed with an SA80A2 fitted with a laser light module. Under the weapon of the second man is a 40mm AG36 grenade launcher. The third soldier is armed with a 5.56mm LSW light machine gun.

A Gurkha instructor of the ANATT has a look at material prepared by an NCO for his next lesson; in this case the subject is anti-ambush drills. In a minute the NCO will start his lesson, and Afghan recruits can be seen waiting for it in the background.

can conduct their missions without being restricted by logistic shortages. The QRF is formed by B Company HQ, two rifle platoons of B Company, a Fire Support Platoon from the battalion's Fire Support Company, and a Swedish infantry platoon. The mission of the QRF is to provide a company-sized formation that can assist any ISAF unit or activity at short notice. The company can be deployed by fixed-wing aircraft, helicopter or by vehicle, and their capabilities range from war fighting to assisting in disaster relief and humanitarian aid missions.

MOT

The forward-deployed eyes and ears of the PRT in Mazar-i Sharif are the Military Observer Teams (MOTs). They are often based in safe houses in remote villages in the provinces, and are looked after by the PRT in Mazar-i Sharif. The MOTs are the link between the PRT on the one hand and local governmental authorities, local representatives of humanitarian aid organizations, and local warlords on the other. The mission of the MOTs is a mixture of monitoring the military and security situation in their region, and to function as a CIMIC element. On the military monitoring side, the MOTs monitor military operations in their AOR. For example, they gather any available information on forces subordinated to local warlords. They also check if local warlords are cooperating with governmental authorities and whether they have handed in their heavy weapons as required under the DDR program. Along with local police forces, the MOTs expend every effort in making the mountainous countryside of northern Afghanistan safer. The teams also coordinate and act as a link during combined ISAF and local police operations.

Gurkhas are skilful shots and are among the top scorers in the British Army. Here a 2 RGR sniper can be seen posing for the camera on a rooftop in Kabul. During operations, snipers can be used as a backup anti-riot force. If necessary, they can take out key targets such as armed rioters or ringleaders. This sniper is armed with a 7.62mm L96A1 sniper rifle made by Accuracy International.

During a patrol in northern Afghanistan, Gurkhas have taken cover behind an abandoned building in hills at the foot of the Hindu Kush. The soldier in the foreground is armed with a 5.56mm LSW light machine gun. Patrolling is part of the daily routine of all Gurkhas deployed to Afghanistan with 2 RGR.

During a patrol, Gurkhas of the KPC hand out ISAF newspapers. The paper is very popular among poor people and often several people read one copy. Due to the fact that a large proportion of Afghans are illiterate, usually children read the paper aloud to their parents.

The CIMIC task of the MOT is similarly challenging. During their usual several-day or weeklong patrols in the mountainous Hindu Kush region, the MOTs gather information about the quality of life of the local population. During talks with town mayors, MOT leaders try to find out where and what humanitarian aid is needed. Required help might range from equipment for a local school through to the repair of a bridge or the building of a well. Once a requirement is identified, the MOT passes the gathered information to the PRT in Mazar-i Sharif, which then takes further action. Among the several MOTs belonging to the PRT in Mazar-i Sharif, five are manned by soldiers of 2 RGR.

Usually a MOT consists of the MOT commander and his deputy, two drivers for the unit's cross-country vehicles, a signaler, a medic and an interpreter. Patrols are carried out either in vehicles or on foot. In some cases the MOTs even hire donkeys from local farmers in order to gain

A patrol of the KPC has stopped at a crossroads in Police District 8. Note the old fortress in the background; it was used by British forces in the First Afghan War (1838-1842). The British Army, as well as its Gurkha units, has seen action in Afghanistan since the mid-19th century. Therefore, Operation Herrick can be seen as continuing history for the Gurkhas.

Showing a presence and thereby deterring criminals and terrorists is a key task of the KPC. Here Gurkhas can be seen on patrol in Police District 8. At least one patrol is on the ground at any one time day or night. For a more approachable appearance the patrolling soldiers have exchanged their combat helmets for infantry green berets with the crossed kukri badge.

Due to climatic conditions, the Gurkhas exchanged their DPM tropical uniforms for desert DPM uniforms. In addition to shirt and trousers, the uniform consists of a desert DPM smock that can be worn in bad-weather conditions. Here a Gurkha can be seen wearing the so-called "Desert Combats" while on patrol in Kabul.

access to areas they would never be able to reach in their vehicles. For the leaders of the MOTs (usually an experienced captain or SNCO), their work is challenging and they often have to operate independently without support from other units. In most cases the success of their mission depends on their willingness to make decisions and use their initiative. In their work the MOTs have the support of the local population and seldom encounter dangerous situations. The best protection for the MOTs is to respect local habits and follow common-sense rules for guests in Afghanistan. Northern Afghanistan is usually quiet and the only threats faced by MOTs might include being caught between competing bands of fighters belonging to local warlords. In mid-May, for example, a MOT of 2 RGR witnessed a gun battle between fighters of General Dostum and General Mallek. The fighting lasted for several hours and continued into the night. The MOT present at the scene observed and reported on the ongoing firefight, which involved approximately 1000 armed men.

The Gurkhas also run the UK National Support Element (NSE) in Kabul, which supports all elements of BRITFOR Afghanistan. Here a Leyland/DAF DROPS 8x6 Medium Load Carrier of the battalion's HQ Company can be seen loaded with supplies that arrived at KIA only hours ago.

"There's nobody quite like us"

"For my soldiers, NCOs and officers, this deployment under Operation Herrick is a welcome challenge," explains LtCol Rigden in his briefing. "Due to the multiple missions, young officers and NCOs often have an enormous responsibility. Several of my men have to work independently and use their initiative to solve problems in the direction of our mission policy. Command and control is a challenge due to the large area we are deployed in, and the fact that the battalion is basically split and based at different locations. Operation Herrick for us is also a nice possibility to get used to operating with several foreign nations," states the battalion commander. "My soldiers conduct joint patrols with German, Canadian and Norwegian troops in Kabul. The QRF consists of a Swedish platoon and often works with German troops in Kunduz. And my CIMIC people have to deal with nearly every nation operating in Afghanistan. The Gurkhas have a long operational tradition in Afghanistan. In past times, Gurkhas fought on the Northwest Frontier and took part in three Afghan Wars (1838-42, 1878-80 and 1919). The 5th Gurkhas distinguished themselves particularly in the Second Afghan War. *Hami jasto kohi chhaina* – 'There's nobody quite like us,' was the unofficial motto of the 5th Gurkhas and my battalion has adopted it in modern times." It is certain that the Gurkhas will continue to use their skills to the hilt, and when redeploying to Brunei, they will be able to say they have served with valor.

In Kabul the Gurkhas can field a couple of Saxon wheeled APCs. Their daily duties involve personnel transportation from Camp Souter to KIA, but in emergencies they can bolster the capabilities of the KPC. The Saxon has a crew of two, consisting of a driver and a commander. In the vehicle's rear compartment there is space for eight fully equipped infantrymen.

The Saxon wheeled APC has a combat weight of 11.7 tons. It is powered by a six-cylinder diesel engine that develops 164hp and allows the vehicle to reach a top speed of 96km/h. The armor of the Saxon protects its crew and passengers against the effects of small-arms fire up to a caliber of 7.62mm. While mainly used for transporting troops, 2 RGR also keeps the Saxon in reserve in case the security situation in Kabul becomes more dangerous.

BRITFOR Afghanistan

British involvement in Afghanistan dates back to late 2001 when U.S. and British Special Forces supported the Northern Alliance in their fight against the Taliban under OEF. In addition, Special Forces searched the country for Osama Bin Laden and his al-Qaeda terrorists. In the next step, British Forces played a major role in establishing ISAF in January 2002. Between January and March 2002, troops of 16 Air Assault Brigade worked as an entry force and prepared for the deployment of other NATO forces in the Kabul area. With ISAF now well in place, the number of British troops under Operation Fingal dropped to a reinforced infantry battalion.

The first Afghanistan Roulemont Infantry Battalion (ARIB) was the 1st Battalion, Royal Anglian Regiment, which deployed for a six-month tour in March 2002. Operational command of the ARIB since then has been exercised by the Kabul Multinational Brigade (KMNB). In October 2004, the name Operation Fingal was changed to Operation Herrick. In mid-2005, some 1050 British troops were operating in Afghanistan under Operation Herrick. Of these, 580 were based in Kabul and some 290 operated with the two British PRTs in northern Afghanistan. The bulk of these troops were provided by the 2nd Battalion, The Royal Gurkha Rifles. Other troops were drawn from the following branches: the Royal Military Police, the Royal Electrical and Mechanical Engineers, the Intelligence Corps, the Royal Army Medical Corps, the Royal Logistic Corps, and the Royal Army Veterinary Corps.

The 200 British troops based in Kandahar belong to a Royal Air Force aviation task group equipped with four Harrier fighter aircraft. Another RAF asset is the so-called "Hilton Crew." Based at Kabul, the unit consists of a single C-130J Hercules transport aircraft with two crews and a ground crew element. Most of the British troops are under the command of ISAF and therefore have the mission of supporting the development of a safe and secure environment for the Afghan population, and the formation of a democratic nation with their own security forces. The Kandahar-based RAF element is under U.S. command and operates within OEF. Their task is to conduct operations in the fight against international terrorism. HQ British Forces Afghanistan is based in Kabul and acts as a national HQ rather than as a tactical HQ for all British Forces operating in Afghanistan.

For patrolling the streets of Kabul, 2 RGR can field a fleet of Land Rover Defender 110 Composite Armour Vehicles. The vehicle protects its crew against the effects of small-arms fire and IED splinters. Note that the vehicles are covered with a "Barracuda" camouflage system, which reduces its heat signature in addition to lowering the visible signature of the vehicle.

The QRF is equipped with several Land Rover Enhanced Protection Vehicles. Here two Land Rover Composite Armour Vehicles can be seen leaving Camp Souter after the QRF was called to an incident.

Drivers of the Saxon wheeled APCs deployed with 2 RGR were drawn from the Queen's Own Gurkha Logistic Regiment. Here one of the drivers can be seen handling a Saxon, which is driven similar to a truck. Note that the Gurkha is wearing a PRR radio that allows him to communicate with the rest of the vehicle crew and his passengers.

Gurkhas on patrol in Police District 9. Illegal house-building activities have taken place in recent years on the hills in the background. This fact may spark trouble in the future as previous owners might reclaim their land. Therefore, the Gurkhas patrol regularly in the area and try to gather as much background information as possible about old and new residents.

Most patrols conducted by Gurkhas in Kabul are joint patrols and they also consist of at least one Afghan policeman. Here such a patrol can be seen passing down one of the several gravel roads in Police District 9. Owing to the rules of engagement during an incident, the policeman is responsible for controlling the situation. If he needs help, the Gurkhas will then provide assistance.

The Gurkhas work together with other assets of BRITFOR Afghanistan. Here a dog handler has joined a KPC patrol in Kabul. He and his dog are assisting in a search of vehicles for hidden explosives and weapons. The dog handler and his dog search the cars while the Gurkhas maintain the vehicle checkpoint, stop the cars and search passengers.

Gurkhas of the QRF of the KPC train for anti-riot operations. During the "Koran misuse demonstrations" in May 2005, the QRF was on standby to support the Afghan Police and ANA in an emergency. But in Kabul the demonstrations remained calm and were well managed. Note the anti-riot gear used by the Gurkhas, which consists of an anti-riot shield, baton, helmet-mounted visor, as well as arm and elbow protection.

In order to disperse a violent mob, the Gurkhas can field an L67A1 riot gun. Smoke and teargas grenades can be fired with this 40mm weapon. In addition, it can be used to fire non-lethal rubber bullets. Here during anti-riot training of the QRF, an L67A1 gunner is seen covered by a comrade using his shield. This tactic allows the gunner to pay full attention to his target while remaining well protected from enemy attack by the soldier with the baton and shield.

Here a soldier of the KPC QRF can be seen in anti-riot gear. Taken during routine training in Kabul, the picture shows clearly the helmet-mounted anti-riot visor, arm and elbow protection, as well as the baton. Note that the soldier is wearing a CamelBak filled with water, as are all his comrades. The CamelBak is an important piece of kit for anti-riot duty as it prevents the soldier from becoming de-hydrated, a condition that can occur quickly in Afghanistan's hot climate.

A joint patrol consisting of Gurkhas from the KPC and Norwegian ISAF troops conducts vehicle searches on one of Kabul's main streets. While the Norwegian soldiers provide security, a Gurkha can be seen searching the minivan. Note that the Gurkha is assisted by an Afghan policeman and an interpreter.

During one of their daylong patrols, a member of a 2 RGR MOT has a chat with an Afghan farmer in a village somewhere in the Hindu Kush. In order to reach remote villages in their AOR, the MOT sometimes use a more regional means of transport - donkeys - in addition to their cross-country vehicles!

Prior to a joint British-Norwegian patrol, a Gurkha patrol commander of the KPC briefs his Norwegian counterpart. Joint patrols are a common activity amongst ISAF troops in Kabul. Joint patrols are a key to interoperability between the different nations that contribute troops to ISAF. Interoperability becomes a key issue when incidents happen along the boundaries of the different nationally controlled AORs in Kabul.

The QRF commander of the KPC reports back to the operations room at Camp Souter on the results of a mission. Note the 7.62mm GPMG mounted on the Land Rover XD 110 WMIK. The QRF is organized into an immediate reaction element and two larger follow-up elements. Depending on the type and size of incident, one, two or all three elements can be deployed. The response times of the elements are also different, though the first QRF asset is able to leave Camp Souter within minutes of being called out.

A Land Rover XD 110 of a KPC patrol makes its way through the busy traffic of Kabul. In order to prevent accidents with merciless local drivers, the Gurkhas have to put all their driving skills to good use. While the traffic shown in this picture is moderate, usually the streets of Kabul are overcrowded with cars and trucks, while every traffic circle is a big traffic jam throughout the day.

After the QRF of the KPC deployed to a suspected incident site, Gurkhas have established a security perimeter. In the meantime, the QRF commander is investigating the incident. The soldier in the foreground is armed with the standard British assault rifle, the 5.56mm SA80A2. Note the Personal Role Radio (PRR) that is part of the Bowman Communication System.

With his 7.62mm GPMG machine gun, a Gurkha provides cover for his comrades conducting vehicle searches at a mobile vehicle checkpoint. The MG is mounted on the weapon ring of a Land Rover XD 110 WMIK.

The Gurkhas belong to the NATO-led ISAF, which at this time operated in and around Kabul and northeast Afghanistan. Note the ISAF badge worn on the upper left sleeve of this Gurkha.

The motto of the Gurkhas of the 2nd Battalion, The Royal Gurkha Rifles, is the former unofficial motto of the 5th Gurkhas: "Hami jasto kohi chhaina" – "There's nobody quite like us." Their professionalism and their language abilities bring the motto to life when the Gurkhas conduct their operations. Here a soldier of 2 RGR can be seen on patrol somewhere in Kabul.

Most of the patrols conducted by the KPC are joint, and therefore consist of Afghan policemen in addition to Gurkhas. The Gurkhas conduct several patrols a week that consist of troops from other ISAF contingents. Here Gurkhas can be seen on patrol with Norwegian soldiers that can be recognized by their 7.62mm G3 assault rifles.

One of the major advantages the Gurkhas have in Afghanistan is that they can communicate with most locals without an interpreter. Most Gurkhas can understand and speak Urdu, a language that is also very common in Afghanistan. Here a patrol commander of the KPC has a chat with a local shopkeeper.

Greek Police EKAM Counter-Terrorist Unit

Juan Pablo Lasterra

Firing a weapon with full assault kit plus a breathing apparatus and a respirator is a difficult task. In this picture we see members of EKAM's NBCR Reaction Team improving their shooting abilities wearing this kind of gear on the range. Note that some of them have a second respirator hanging from their belts. (EKAM NBCR Team)

The 2004 Olympic Games in Athens resulted in important human and material growth for EKAM, the elite unit of the Hellenic Police. This has had a very positive effect on the quality of the group.

On a suffocating morning in June we had the opportunity to witness several training sessions of an EKAM team in Athens, with the purpose of providing photos for this article. We also visited the EKAM base in order to learn something about the unit's history, structure, training, missions and equipment. The exercises consisted of several series of descents from a training tower using different techniques each time. Among them were standard and Australian rappelling, abseiling, and head-down descents.

We also visited different structures representing the interior of airplanes and buildings especially constructed to meet the unit's needs. The agents can train inside using live fire or simunitions. The marksmen also had their place in the exhibition as they deployed some of their sniper rifles fitted with the latest-generation optical sights like the diurnal/nocturnal AN-PVS10 scope. The final touch to the first part of our visit was provided by two men of the unit with their Glock 21 pistols chambered in .45 ACP, as EKAM is one of only a few European units to routinely use this powerful and effective cartridge in their sidearm. While an agent held two balloons with his hands just a few centimeters from his head and a third balloon with his teeth, his companion a set distance away burst them with three accurate pistol shots. Immediately after that, the first agent spun around and emptied his own Glock magazine with extreme precision into different targets placed on the range.

After these demonstrations we went to the unit's headquarters with the purpose of seeing more equipment currently in service and to have an interview with Officer George P., who is in charge of one of EKAM's assault groups.

EKAM operators equipped with Dräger breathers and respirators jump from an unmarked civilian van during training in an intervention scenario where a nuclear, biological, chemical or radiological (NBCR) weapon of mass destruction (WMD) has been used by terrorists. (EKAM NBCR Team)

The EKAM NBCR team assaults a house where simulated terrorist targets are supposed to be hiding a WMD device. (EKAM NBCR Team)

The last member of this NBCR team is equipped with a full protective NBCR suit. This gear completely protects the officer from any known NBCR agent, but after some minutes of use it will be very hot inside the suit. A very important point is that, like any gasmask filter, these protective suits have an expiration date after which they must be withdrawn from service. (EKAM NBCR Team)

EKAM: The Greek answer to terrorism and crime

Eidiki Katalstaltiki Antitromokratiki Monada (EKAM) means Special Counter-Terrorist Unit. According to the commander, "The unit was created in 1984 by the union of the Greek Gendarmerie and Urban Police into a single organization. Each one of these law enforcement agencies had possessed an intervention unit since 1978, named Special Operations Group (TEIDA) and Special Operations Platoon (DEIDA) respectively. Thus both special teams were united to form the present-day EKAM. The process was not easy because both units had very different training, armaments and roles."

During its early years, EKAM was composed of 150 agents, a figure that has recently been increased with the arrival of another 50 members due to the security exigencies demanded by the holding of the Athens Olympic Games in 2004. All the officers dedicate themselves fulltime to EKAM. In these uneasy times in which most similar European special units have serious problems in enlarging their workforce, we asked the commander about EKAM's formula for success. "It was also difficult for us to increase our personnel. In the four years preceding the Games we made several annual requests looking for that number of new personnel. For each request we received 50-60 candidates, but only one-tenth of these could enter EKAM. During the pre-Olympic period, EKAM also received great help from European counterparts. After the Games, we realized the benefits of this long process, not only because of the arrival of new equipment, but because we now have much more flexibility in programming tasks and training. Overall, we are prepared to solve a much greater array of missions with success than we were five years ago."

EKAM is commanded by a lieutenant colonel and it is divided into six intervention groups, each headed by a commander. Four of the groups are based in EKAM's headquarters in the capital while the other two are on permanent and uninterrupted monitoring duty in the main airports of the country - Athens and Thessalonica. In order to enter the unit it is necessary to be aged less than 28 years old in the case of sergeants, or less than 30 for officials. Selection tests are not held annually but, as we have seen, are held in response to a requirement for new members. Usually 15% of the candidates manage to pass the first selection course lasting 52 days.

All the members of the unit are trained as assaulters, although each agent

The bulky NBCR gear protects the officer from all known aggressive agents, but as it is completely sealed, the gun belt and other essential equipment have to remain outside the suit. If the officer had to intervene in a real incident where NBCR agents had been dispersed, he would need to decontaminate all his equipment. If he used a leather belt, that item would have to be destroyed since leather absorbs agents. (EKAM NBCR Team)

The use of WMD by a terrorist group during the Athens Olympics in 2004 represented a threat that was taken very seriously by the Greek authorities a long time before the Games. Therefore, an NBCR Reaction Team was created within EKAM and was trained and equipped to perform all of the unit's usual missions but in a contaminated environment. This team also intervenes if there is a suspicion that a terrorist group is in possession of WMDs inside Greece. (EKAM NBCR Team)

An EKAM NBCR Reaction Team member, wearing full assault kit and protective gear, trains with his FN P90 PDW at the shooting range. (EKAM NBCR Team)

can specialize in a particular field of work according to the needs of the unit: marksmanship; breaching techniques using explosives, hydraulic or mechanic tools; combat diving; static-line parachuting; skiing; battlefield paramedic; and intervention in Nuclear, Biological, Chemical or Radiological (NBCR) contaminated scenarios in which a Weapon of Mass Destruction (WMD) has been used. In fact, one of the hypotheses the Greek authorities worked on before the Athens Olympics was a possible terrorist attack using a similar type of weapon. For that reason, EKAM stood up an NBCR Reaction Team a year before the Games. This team was formed of several operators trained and equipped to carry out the usual range of missions but in an NBCR environment.

In addition to the facilities mentioned at the beginning of this article, EKAM frequently trains in places such as Athens Airport, the port of Piraeus, city subway and train systems, and various structures built for the Games. Their relationship with similar foreign units is very fluid. In fact, nearly all the agents have trained outside Greece. The closest links are kept with the following units: Spanish GEO, French RAID and Lyon GIPN, American FBI's HRT, Scotland Yard's SO19, Berlin's SEK, and Austria's Cobra, without forgetting some training and interchanges with U.S. Army Special Forces and, of course, with the Greek military Special Forces community. The Greek Armed Forces cooperate with the unit by supplying C-130 Hercules aircraft and CH-47D Chinook and Aerospatiale Super Puma helicopters for aerial transport needs, as well as SH-60 Sea Hawk and Super Puma choppers for assaults in which "fast-rope" techniques are used.

EKAM's shoulder badge.

A portrait of an EKAM NBCR Reaction Team member wearing full assault kit and a bulletproof vest, a Dräger breather, and a respirator. Note that his assault boots are also protected by special covers. (EKAM NBCR Team)

An EKAM assault team approaches the unit's "kill house" during a hostage rescue drill in Athens.

High-risk missions

The missions assigned to EKAM are hostage rescue, the arrest of dangerous criminals, high-risk VIP escorts, collaboration with Hellenic security forces during antiterrorist operations, countering organized crime, and general rescue operations that might include collaboration with the fire services. The unit operates directly under the command of the Hellenic Police Chief and it has the prerogative to act in all national territories and even to go abroad - for example, to protect Greek embassies or diplomats in countries experiencing conflict. Among the numerous missions in which EKAM has participated since 1984, Commander George P. remarked that 50 of them have been catalogued as very high risk. He emphasized the following ones:

- 1999. An Albanian citizen hijacked a busload of passengers in the northern county of Florinas. He was armed with a belt filled with explosives and two hand grenades with their safety pins pulled out. After more than 36 hours of fruitless negotiation, the Greek authorities decided to storm the bus. An EKAM marksman initiated the action by shooting the kidnapper dead with his silenced sniper rifle, while two assault teams simultaneously penetrated the interior of the bus and managed to catch the grenades before they fell to the ground.
- 1999. EKAM helped civilian rescue teams to search for buried victims after an earthquake that devastated some of Athens' neighborhoods on 17 September. The quake killed dozens of inhabitants and destroyed hundreds of buildings.

Armed with the ubiquitous H&K MP5A5 sub-machinegun, in this case fitted with a Surefire flashlight and an EOTech holographic sight, these two EKAM officers are ready to storm a house during a counter-terrorism exercise.

An assault team from EKAM storms the "kill house" during a counter-terrorism exercise. Note the full protective ballistic vests worn by all the officers.

The unit's "kill house" is constructed to permit changes and rearrangements in the interior by using mobile panels. Many of the exercises are carried out using either live ammo or simunitions.

EKAM fields different tactical accessories for all its weapons currently in service - for example, these FN P90 PDW weapons in 5.7 x 28mm caliber fitted are with laser pointers and flashlights.

An EKAM officer aims his H&K MP5A5 during a hostage rescue drill in the unit's "kill house". Note the impact marks on the walls from bullets fired in previous live-fire exercises.

- 2002. A large operation was mounted against the Greek November 17th terrorist group responsible for multiple assassinations, bombings and hold-ups. EKAM contributed to the final 19 arrests by keeping in custody in hospital some of the terrorists hurt during the assaults, and guarding detainees being transferred to penitentiary centers.
- 2002. A young person kidnapped his fiancée and threatened to kill her with a knife. An EKAM operator used a Taser M26 electrical pistol to neutralize him without employing lethal force. According to the commander, the Taser option has been used successfully in numerous interventions by his unit.
- 2003. A hijacker took 194 hostages in a Turkish Airbus flying from Istanbul to Ankara. Threatening to blow the plane up with an explosive-packed belt, the aircraft landed at Athens Airport to refuel. After some hours of negotiations, EKAM assaulted the airplane and arrested the aerial pirate in the cockpit. The threat about the belt turned out to be false.
- 2004. EKAM provided security during the Olympic Games by protecting the main venues and tourist centers of Athens, and escorting several delegations of athletes. The unit's deployment included elite marksmen patrolling in helicopters.
- 2004. Two Albanians armed with shotguns hijacked a coach with 23 passengers in Athens. EKAM surrounded the vehicle with assault and sniper teams. After 19 hours of negotiations, the kidnappers surrendered to the police.

Latest generation of tools of the trade

In its armory, EKAM holds weapons in a varied range of calibers. Without doubt, we could easily say that EKAM has been one of the pioneers in introducing some weapons into Europe. Here is a list of weapons in EKAM's arsenal.

- Pistols: Glock 21 in .45 ACP equipped with an ITI M6 light; SIG Sauer P229 in .357 SIG; FN Five Seven in 5.7 x 28; and Ruger GP100 revolver in .357 Magnum. The non-lethal Taser M26 electric pistol is also employed.
- Sub-machineguns: H&K MP5 in 9mm Parabellum in all its variants, equipped with holographic EOTech sights, Surefire tactical lights, and B&T tactical hand-guards; and FN P90 PDW in 5.7 x 28 equipped with laser pointers and lights.
- Assault rifles: AK-47 in 7.62 x 39; M16A2 and M4A1 carbines in 5.56 x 45, equipped with various scopes and aiming aids.
- Sniper rifles: in 7.62 x 51 caliber, EKAM uses Sako SSR Mk3 models with integral silencer and 6 x 44 scope; H&K G3SG1 with Carl Zeiss 1.5-6 x 42 scope; Accuracy International AW with Schmidt & Bender 3-12 x 50 scope; and Knight's Armament/Stoner SR25 with Leupold 3-12 x 50 optics. In .338 Lapua Magnum, the chosen rifle is the Accuracy International Super Magnum with Schmidt & Bender 3-12 x 50 scope. All these sniper rifles can also use the latest-generation AN-PVS10 diurnal/nocturnal scopes. The ammunition used is Match and AP types.
- Machine guns: FN MAG in 7.62 x 51.
- Shotguns: Remington 870 in 12-guage.

Officer George P. admitted that, "The budget granted to EKAM before the Athens Games allowed us to have a very complete set of equipment, all that was necessary. In fact, at the present time we don't plan to increase it for a period." The list of gear and equipment is very extensive and it is impossible to summarize in a single article. We can simply say that the basic assault kit is typically composed of non-flammable Nomex coveralls, a Kevlar helmet, a gasmask and individual communications equipment (made in the UK by Davies), a tactical Greek vest by EVO, English Warrington Pro assault boots, and American Blackwater pistol holsters. The unit has diverse "ghillie" camouflage suits for snipers, closed- and open-circuit diving breathers, thermal and infrared detection cameras, plus extensive equipment specifically conceived for the diverse intervention scenarios in which the unit may be required to intervene.

The NBCR Reaction Team has Dräger breathing apparatus and respirators and full protective suits. The most common vehicle in use by EKAM is the 4x4 Jeep Grand Cherokee, although EKAM also uses vehicles in "civilian" camouflage as "Trojan horses". Zodiac boats are also used for interventions in the sea or in rivers.

EKAM's "kill house" allows the use of live rounds during training. The instructors can follow the progress of officers from above the rooms during tactical training.

Portrait of an EKAM operator armed with an FN P90 PDW. Note the communications microphone, full protective ballistic vest, and the unit's badge worn on the Nomex suit.

Portrait of two officers from EKAM, both armed with the FN P90 PDW in 5.7 x 28mm caliber and fitted with laser pointers and flashlights.

A close-range shooting exercise performed in the training area that EKAM uses for pistols, sub-machineguns and shotguns. EKAM operators usually carry a Glock 21 in .45 ACP as their "on-duty" sidearm – it is equipped with an ITI M6 light. This caliber is not found very often among European counter-terrorism teams, so it is supposed the Glock 21 was adopted following advice from the American FBI HRT and U.S. Army Special Forces.

A demonstration is performed in order to build and reinforce the confidence of comrades. While an operator holds three balloons close to his head in his hands and mouth, a colleague fires his Glock 21 pistol at close range, bursting them. Immediately after that, the officer turns towards a target and empties his own Glock 21 magazine at close range.

In order to practice hostage rescues in airplanes, EKAM had a full-scale model built in their training area. Of course, there's nothing like practicing on real planes, so the unit also trains on all kinds of commercial aircraft. A team of assaulters waits for the green light to storm an airplane during an exhibition of their abilities.

The breacher of an assault team enters the cabin of EKAM's airplane model. Note the bullet pockmarks on the walls - this training aircraft permits the use of simunitions and other devices.

A pair of assaulters armed with FN P90 and H&K MP5A5 sub-machineguns clears the aisle of the simulated aircraft during a demonstration at their training area.

The 12-gauge Remington 870 is perhaps the most famous pump-action shotgun in use by police and military around the world. It was introduced by Remington in 1950 and is capable of firing buckshot, slugs and special-purpose munitions.

The "less-than-lethal" advanced Taser M26 electric pistol has been used by EKAM in cases where the use of lethal force was unnecessary to subdue a suspect. The pistol uses compressed nitrogen to project two small probes out to a maximum range of 10.6m (depending on the cartridge) at a speed of more than 50m/s. Both probes are connected to the Taser by insulated wire. An electrical signal is transmitted through the wires to where the probes make contact with the body or clothing, resulting in an immediate loss of that person's neuromuscular control and the ability to perform coordinated action for the duration of the impulse, regardless of mental focus, training, size, or drug-induced dementia of the suspect. The M26 also has a built-in laser sight.

The FN Five Seven pistol shoots the same SS190 5.7 x 28mm round as the FN P90 PDW. It has a 20-round magazine and weighs only 760g loaded. Available only for military and police units such as EKAM, the SS190 5.7 x 28 round is credited as being able to penetrate a standard PAGST vest at 300m or a standard CRISAT (Kevlar and titanium) vest at 100m, both without over-penetration.

Three EKAM officers are ready to rappel down the tower during a demonstration for "Special Ops Journal". Athens is in the background, swathed in heavy pollution.

Three EKAM officers rappel down the training tower. They are using a window to practice storming a house from the exterior.

Protected only by a Pro-Tec helmet, an EKAM officer starts climbing the unit's training tower without any harness. In a real operation in a building the officer is required by circumstances to climb this way and that is the reason why safety accessories are omitted here in training.

The training tower of EKAM is six stories tall and is used for practicing all kinds of rappels and descents. It is also, as on this occasion, used to practice different climbing techniques. Note that he has no harness or any other safety device other than his Pro-Tec parachute helmet.

The Omer 1 descent lever used by EKAM is made from three pieces of milled aluminum (body, control lever and front plate) and several minor stainless steel components. The system allows the officer to control each moment of the descent, even to stop it automatically if need be. It thus allows him the free use of both his hands.

An EKAM officer armed with a Glock 21 pistol shows one of the different entry techniques used by his unit to storm a house.

An EKAM officer, Glock 21 pistol in one hand and Omer 1 descent lever in the other, prepares to rappel down the training tower.

Two perfectly camouflaged EKAM snipers in the area surrounding the unit's training ground. In this picture we can see some of the facilities used during exercises (e.g. the artificial aircraft cabin, a pistol and sub-machinegun shooting range and a "kill house").

Two views of an EKAM tandem rappelling head-down on their training tower. All the policemen of the unit are trained to use different rappelling techniques depending on the required tactical situation.

An EKAM operator proudly poses under the emblem of his unit while holding a 7.62 x 51 Sako SSR Mk3 sniper rifle. Made in Finland, this weapon has an integral silencer and a 6 x 44 scope. Note the officer is equipped with a full assault kit rather than proper sniper gear, as he is posing for photographic purposes only.

This rappelling technique has multiple uses. It can serve to evacuate a wounded comrade or civilian from the top of a building. It also allows an agent to cover his comrade from any threat during the descent.

A pair of EKAM snipers armed with an H&K G3SG1 semi-automatic rifle fitted with an AN-PVS10 diurnal/nocturnal scope, plus an Accuracy International AW bolt-action rifle equipped with a Schmidt & Bender 3-12 x 50 scope. Both weapons use the .308 round (7.62 x 51mm NATO cartridge). EKAM snipers also use the Super Magnum model made by the same British manufacturer, a weapon that uses the powerful .338 Lapua Magnum round.

Many of the sniper rifles in use with the EKAM, like this H&K G3SG1, can be provided with an AN-PVS10 diurnal/nocturnal scope. The AN-PVS10 provides the sniper with the capability of acquiring and engaging targets during low- and high-ambient light conditions. For nighttime "business" the scope uses third-generation image intensification technology, while for day use a direct-view system is employed. The main drawback of the device is its weight and the fact that it is difficult to hide its form.

EKAM snipers are trained to use alternative shooting positions to the classic prone position, as in some scenarios it is not possible to use the latter. Leica Vector rangefinders combine binoculars, an eye-safe laser rangefinder, a digital compass and an inclinometer in a single compact device. They measure distance, azimuth and elevation of targets as far as 10km away. They also display horizontal distance, height difference and azimuth between two remote objects. In other words, it is a really useful tool for a sniper spotter!

Wielding an Iron Sword
NATO Reaction Force in Norway

Carl Schulze

This picture shows uniforms, weapons and a vehicle of the 1st Light Reconnaissance Squadron. Normally in such an environment, soldiers and vehicles are totally camouflaged and cannot be seen even at a distance of five meters!

NATO Response Force under Dutch Command

The NATO Response Force (NRF) is a coherent, high-readiness, joint, multinational force package. It uses advanced technology, and is interoperable and sustainable. Indeed, the NRF is the key catalyst for focusing on and promoting improvements in NATO's military capacities.

In September 2002, the U.S. Secretary of Defense, Donald Rumsfeld, put forward a proposal to create a NATO rapid reaction force. The launch of the NATO Response Force initiative was announced a couple of months later at the Prague Summit in November. During this summit, all NATO members agreed transformation of the Alliance was needed to face the threats of the 21st century. For this purpose, a rapid decision-making process and a rapidly deployable military force were required. Another reason for the change was developing the armed force structure – for example, setting up rapidly deployable headquarters for the army and navy. The changing relationship within the European Union (EU), and closing the military gap between the USA and its allies, were also factors in the decision.

A Mowag Eagle is seen in a typical Norwegian environment. The Mowag Eagle I is the equivalent of an armored M1097 HMMWV. It has a top speed of 125km/h and a gross vehicle weight of 5100kg, with a payload of 1200kg. The vehicles are not very old but they have seen extensive service with ISAF, SFOR, KFOR, and in Iraq, so they are now beginning to wear out.

Portrait of a well-equipped Danish reconnaissance soldier. One member of the unit is the Danish judo champion! He is wearing standard Danish camouflage designed using computers, and is armed with a C7 rifle, the Canadian version of the M16. Here it is fitted with a laser pointer and an ELCAN scope. Surefire is also used on peacekeeping missions.

In addition to the decision to review NATO's command structure so as to enable it to respond and deploy more rapidly, a major breakthrough in regard to military capabilities was achieved in Prague. This involved the setting up of the NRF. Since Prague, the NRF concept has been developed continuously. Nowadays it comprises approximately 20,000 troops and it can be deployed within a period of 5-30 days. An instance of when this might be necessary could be defending the Alliance in the initial phase of a crisis response operation, or in the event of an unexpected escalation of a conflict.

The NRF is composed of national force contributions that rotate through periods of training and certification as a joint force, followed by an operational "stand-by" phase of six months. The NRF is composed of a Land Component, Maritime Component, Air Component, and Special Forces Component that are led by a Deployable Joint Task Force.

The NRF is a force package, so it is tailored as required to the needs of specific operations, and it is able to move quickly to wherever it is needed. The NRF can carry out certain missions on its own, or it can serve as part of a larger force while contributing to the full range of NATO military operations. The units belonging to the force package can deploy within 5-30 days and it can sustain itself for up to a month, or longer if it is appropriately re-supplied.

A patrol prepares for action deep in a forest. The 1st Light Reconnaissance Squadron has operational experience in extremely cold climates. Note the effectiveness of the Danish camouflage in this forest environment.

NRF Toolbox – Composition of the NRF Brigade

43 (MN [Multinational]) Mechanized Brigade of the Royal Netherlands Army is one of the main parts of the Land Component Command as part of the NRF's fourth rotation. As of 14 January 2005, 43 MN Mechanized Brigade was in the stand-by phase. It consists of 4000 soldiers from five European countries and has a great variety of combat, combat support, and logistic units. It is led by the staff of 43 (NLD) Mechanized Brigade, reinforced by personnel from the other contributing countries.

The NRF-4 Brigade combat unit comprises: a Danish light reconnaissance company, a Dutch mechanized infantry battalion, the Norwegian Telemark Battalion, a Dutch infantry air assault battalion, and a German airborne infantry battalion group. Combat support is provided by French field and air defense artillery, with a contribution also coming from Spain. The Netherlands also provides an armored engineer battalion. Together, Germany and the Netherlands take care of logistics. All these units communicate via the Dutch TITAAN system.

The medics of the Danish Light Reconnaissance Platoon are particularly well trained. Here they simulate treating a casualty while under fire. Two medics give first aid to a wounded reconnaissance soldier as they are covered by other members of the team.

Danish reconnaissance soldiers in action with a C7 rifle. In NRF-4, the Light Reconnaissance Squadron is organized with an HQ platoon (61 men), an engineer team in charge of EOD (Explosive Ordnance Disposal) with eight men, three reconnaissance platoons of 23 men each, and a National Support Element.

Combat Units

The Dutch mechanized infantry battalion, 44 (NLD) Mechanized Infantry Battalion, and the Norwegian Telemark Battalion provide the maneuver power of the NRF Brigade. 44 (NLD) Mechanized Infantry Battalion consists of elements from 44 (NLD) Armored Infantry Battalion and 42 (NLD) Tank Battalion. The battalion is made up of a staff HQ, mortar platoon, logistics elements, and two mechanized teams. There is no need to give further details about the Norwegian Telemark Battalion, since this unit was the subject of an earlier separate article.

44 (NLD) Mechanized Infantry Battalion is a combination of mechanized infantry and tank platoons. The unit will eventually consist of two mixed teams, an HQ and service support company, a mortar platoon, and a reconnaissance platoon. It will also contain various logistics elements. Depending on how and where the NRF deploys, the mixed teams may be configured differently. The basic composition at present is two tank platoons, two mechanized infantry platoons, and one antitank platoon. Both teams are capable of carrying out both regular and crisis response missions. They are both well trained for operations in urban areas and in riot control.

The air assault infantry element is called 12 (NLD/DEU) Air Maneuver Battle Group (BG), and it contains the Dutch 11th Luchtmobiele Brigade (Airmobile Brigade). This BG conducts air assault operations for the NRF Brigade. The infantry battalion is made up of two infantry companies, a staff company, an air assault 120mm mortar platoon, an air assault air defense platoon, an air assault engineer platoon, and a logistics detachment.

Each fire team has a C9 Light Support Weapon fitted with a heavier barrel. The C9 weighs 5.8kg empty, and 6.2kg loaded. Its rate of fire is 625 rounds per minute. The squadron fields seven such weapons.

In addition to the Dutch infantry battalion, there is a German airborne infantry battalion - 373 Fallschirmjäger Battalion. As it was not present at Exercise Iron Sword 2005, we will not present details about this unit here.

1 (DNK) Light Reconnaissance Squadron provides the eyes and ears of the brigade. It consists of an HQ platoon and three reconnaissance platoons. They move by means of the MOWAG Eagle light armored vehicle. In addition to reconnaissance tasks, the company specializes in protecting and escorting convoys.

Combat Support Units

A French-led Artillery Battle Group comprising field artillery and air defense artillery, provides fire support for the brigade. This battalion is equipped with a battery of eight tracked AU-F1 155mm self-propelled howitzers. Furthermore, the battalion consists of a battery of four MLRS rocket launchers, an airborne 120mm mortar battery, and an air defense battery. Spain provides 24 Air Defense Battery from Valencia, which is equipped with the Mistral missile launcher.

During the NRF-4 rotation, 11 (NLD) Armored Engineer Battalion will

A well-camouflaged soldier armed with a C9. To show the weapon to readers, the weapon has not been camouflaged. Each patrol consists of two Mowag Eagles with three reconnaissance soldiers in each. The platoon has three reconnaissance patrols and one command patrol.

provide and coordinate the engineer capability. The battalion comprises a staff element and 43 (NLD) Armored Engineer Company (Heavy). To help NRF-4 cross various obstacles during Exercise Iron Sword, 101 (NLD) Engineer Battalion, with 800 men in six companies, was dispatched to Norway. The six companies are as follows: 101st NBC Company with Fuchs 6x6 vehicles, 102nd, 103rd and 104th Engineer Companies, and 105th Bridge Company. Norway is famous for its fjords that represent important obstacles, but in military terms, these were not the only obstacles encountered in Telemark province where Iron Sword 2005 was held.

101 (NLD) CIS, the new Communication and Information System Battalion of the Royal Netherlands Army, provides communications. Throughout NRF-4, TITAAN is the main means of communication, this system allowing all units to communicate with each other and to freely exchange information.

Combat Service Support

German and Dutch logistic units take care of logistics for the NRF Brigade. 43 (NLD) Maintenance Company and 43 (NLD) Supply Company, along with a German logistics company - 272 (DEU) Logistics Company - are responsible for the brigade's logistical needs.

Headquarters

The Brigade HQ commands all brigade units. This HQ consists of staff personnel of 43 (NLD) Mechanized Brigade, and they are augmented by personnel from other contributing nations. It is supported by 43 (NLD) HQ Company, while 43 (NLD) Staff Guard Company provides protection. This latter unit, consisting of personnel of the organic reconnaissance squadron of 43 (NLD) Mechanized Brigade, not only protects the HQ but also protects convoys.

A Mowag Eagle I. A total of 36 Eagle I vehicles were delivered to the Danish Army to replace their upgraded M41 Walker Bulldog light tanks. Some Eagles have an MBK 2 turret mounting an MG-3 machine gun. Two of these vehicles can be loaded in a single C-130 Hercules of the Royal Danish Air Force.

Intelligence images of the battlefield are acquired by NRF-4 via 103 (MN) ISTAR Battalion, a multinational ISTAR (Intelligence, Surveillance, Target Acquisition and Reconnaissance) unit consisting of Dutch, Romanian, Polish and Norwegian elements.

Dutch Weapons

In 1984, Canadian Forces adopted a new generation of 5.56mm caliber weapons for all its services. The selected weapons were based on the M16 design and incorporated some of the features of the A1 and A2 versions of this famous American assault rifle. The C7 retains the fully automatic firing mode rather than burst firing, as well as a single-shot mode. The rifle has a case deflector to permit left-handed firing, and the barrel is a cold-forged, chrome-lined design rifled to accept both SS109 and M193 type ammunition. This same family of Diemaco weapons has been selected by the Dutch army to equip its soldiers.

C7 Assault Rifle Specifications

Characteristics:	cartridge 5.56 x 45mm
Operation:	direct gas, selective fire
Locking method:	rotating bolt
Feed:	30-round box magazine
Weight:	3.3kg (empty), 3.9kg (loaded)
Length:	1.02m
Barrel length:	510mm
Rifling:	6 grooves rh; one turn in 178mm
Sights:	fore, post rear, 2-position aperture

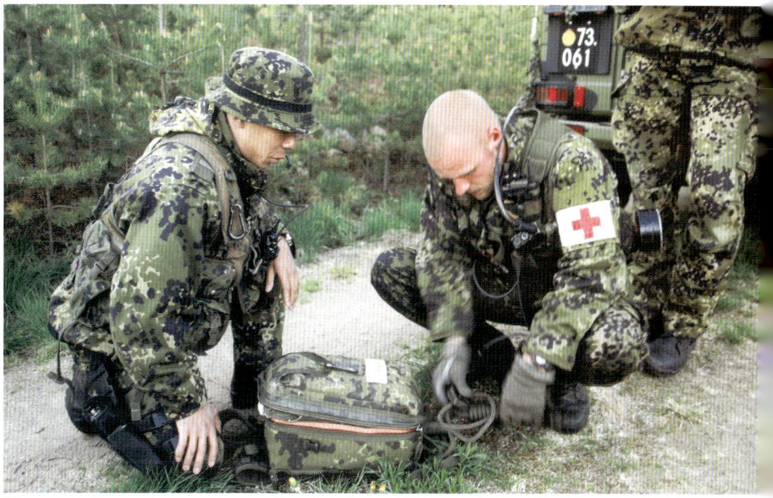

The medics are among the best in Denmark, and they can be deployed with deep long-range patrols. It takes six months to train a medical team. Note their medical packs. The Medical Platoon contains 17 men. Two M113G3 vehicles serve in this platoon attached to the HQ Platoon.

A Fokker 60 of 334 Squadron of the RNLAF drops a troop of reconnaissance pathfinders over the small airport at Rena. These troops belong to 11 (NLD) Air Maneuver Brigade. The parachute is the French-manufactured Air Azur 626-29C, while the reserve chute is the S11.

Normally such a parachute drop takes place at night. Its purpose is to infiltrate reconnaissance pathfinders who will then clear an area in which to launch the subsequent airlift.

Once the airport's control tower has been captured and the perimeter secured, the pathfinders' leaders will give the green light for the airlift to proceed.

Once on the ground, the pathfinders take control of the small airport in a fast and brutal assault. Weapons seen here are the Diemaco C7A1 assault rifle and C8A1 carbine. Note that all soldiers have knee and elbow protection.

Six C-160 Transall aircraft of the Luftwaffe are used to airlift in the 12th Infanteriebatalijon "Regiment van Heutsz", which is one of three famous air assault battalions of the Dutch Army. Reinforcements also comprise the 11th Mortier Compagnie (Mortar Company) and 11th Luchtverdedigings Compagnie (Air Defense Company).

Airlifting NATO infantry units into a small airport symbolizes the end of the Cold War. The Cold War was characterized by troops positioned in static positions confronting the threat of a large mechanized attack by the Warsaw Pact along a linear front. NATO is now experimenting with new tactics where troops are deployed far from home territory, just as occurred with the French in recent airlift operations in Africa.

Reinforcements are permanently inserted into the airmobile bridgehead, and here some light vehicles are airlifted in by a Luftwaffe Transall. Mercedes G 4x4 vehicles armed with MAG machine guns take up position.

After being airlifted in, airmobile soldiers of the 12th Infanteriebatalijon "Regiment van Heutsz" occupy positions around the airport to extend the perimeter established by the pathfinders. The picture is interesting because it shows different infantry weapons such as the MINIMI, Accuracy International sniper rifle and C7.

The Accuracy international AW sniper rifle fires 7.62 x 51mm (301 Winchester) rounds. Its weight is 6.4kg, and it has an overall length of 1.18m (the barrel length is 660mm).

Luchtmobiele Brigade snipers and sharpshooters field the Accuracy International AW 7.62mm sniping rifle, which is also used by a large number of other NATO countries.

After the capture of the airport at Rena, airmobile soldiers of the 12th Infanteriebatalijon "Regiment van Heutsz" launch an amphibious raid to attack a terrorist camp.

These pictures show the Dutch airmobile soldiers familiarizing themselves with inflatable boats provided by the 101st Engineer Battalion before the raid.

After disembarking from the rubber boats, soldiers of the first wave take cover on the riverbank to assume protection for the second wave. It was a good opportunity to see the weapons of the Dutch airmobile soldiers – C7A1 assault rifles, M203 40mm grenade launchers, and MINIMIs fitted with a Kite Mk VI sight.

Crewmen of a Patria XA-188 guard a bridge and observe what is happening on the other side of the river. Two of these large vehicles are supplied to each company. The Netherlands has 70 Patria XA-188 vehicles that are used solely for peacekeeping missions.

Note the famous Indianhead badge on the front of this Patria XA-188, as well as the large wire cutter post.

A Patria XA-188 seen in the Norwegian countryside. After troops of the airmobile units are airlifted in, these vehicles can provide fast and light armored transport for the Quick Reaction Force (QRF) of the battalion.

Further views of a Patria XA-188 attached to the 12th Infanteriebatalijon "Regiment van Heutsz" air assault battalion.

The LSV (Luchtmobiel Special Voertuig) is a French-manufactured light vehicle made by the Lohr Company. It is used by all ground units of the 11th Dutch Airmobile Brigade to improve mobility once units have been airlifted in.

Today, 175 LSV vehicles in different variants are used by the brigade. Here is an LSV-AD (Allgemeine Dienst – "General Purpose") variant of the 12th Infanteriebatalijon "Regiment van Heutsz".

Some LSV-AT vehicles are seen ready to take up an ambush position on a main road. These vehicles are perfectly suited for light airmobile infantry combat, and they could inflict damage on even the largest armored formations if properly employed.

This LSV-AT vehicle is carrying AT-4 light rocket launchers and JIL missiles.

Reconnaissance divers of the 101st Engineer Battalion are used to perform reconnaissance of the riverbank in preparations for a river crossing. Weapons are the standard C7 used by the Dutch Army.

This photo shows the launching of a ribbon bridge pontoon by the 105th Company of 101 (NLD) Engineer Battalion at a crossing site previously reconnoitered by divers. The company has one reconnaissance diver platoon and two bridging platoons.

Each bridging platoon has three rafts and the assembly of these allowed two bridges to be formed in less than 90 minutes.

Dashing from the depths of a boreal forest where it was camouflaged, a MAN truck deploys its "iron carpet" at the site of a river crossing.

This picture shows the MAN truck is a good swimmer.

After their arrival, the soldiers quickly debussed from their AIFV YPR-765 and took up a position in a single line. The soldiers' appearance is very similar to their airmobile colleagues, being armed with C7A1 rifles, some fitted with ELCAN sights.

A scenario similar to those encountered during peacekeeping missions is played out with great realism. Some ethnic Trons have attacked a school and taken some hostages. The QRF of Bravo Team of 44 (NLD) Mechanized Infantry Battalion has just arrived on the scene.

The Dutch soldiers are needled by the rioters. They keep cool thanks to their training, and do not respond to the provocation. Actually, all NATO troops are trained to face this kind of situation, one that has actually been encountered by the French and Danes in Mitrovica.

After half an hour of harassment, the Dutch commander decides to use force to disperse the rioters. For this purpose he uses four AIFVs of the platoon to rush in ranks towards the "bad boys". The psychological impact of tracked armored vehicles is still one of the best deterrence tools.

Here, an AIFV of 44 (NLD) Mechanized Infantry Battalion is seen in action. It is a basic YPR-765 PRI infantry version. It has a crew of three and space for seven infantrymen. The main armament consists of a dual-fed Oerlikon 25mm KBA-B02 cannon, which has 180 ready rounds and a further 144 rounds in reserve. Mounted coaxially to the left of the cannon is an FN 7.62mm MG with 230 ready rounds and a further 1610 rounds stowed in the hull.

A YPR-765 PRAT (Pantser Rups Antitank) is seen in an ambush position alongside a forest track. From here it can quickly engage a target and then take cover. Note the soldier's green beret with a red flash, which is the color of 44 (NLD) Mechanized Infantry Battalion. A normal company consists of three platoons of four vehicles, plus an antitank platoon.

The first AIFVs entered Dutch Army service in 1982. A total of 2079 vehicles in 15 different variants are now in service. An AIFV YPR-765 of 44 (NLD) Mechanized Infantry Battalion reaches a staging position during Exercise Iron Sword 2005 in the depths of a forest. The vehicle has a weight of 13,687kg and a road speed of 61.2km/h.

This YPR-765 PRAT is fitted with an Emerson TOW launcher, the same as that fitted on the M901 of the U.S. Army, as well as the USMC's current LAV-AT. Dutch vehicles are also characterized by a battery of six smoke grenade launchers mounted on the front of the hull.

A YPR-765 PRVA cargo carrier and a YPR-765 PRBRG recovery vehicle (note its crane) pass a Leopard 2A6. The YPR-765 PRVRA, a cargo version of the AIFV with a crew of three, has a combat weight of 13,400kg. It is armed with a 12.7mm pintle-mounted MG on a standard M113 cupola.

A Leopard 2A6 of 42 (NLD) Tank Battalion is seen during a road move at the beginning of the exercise. This version of the famous Leopard 2 is immediately recognizable by its longer 120mm barrel.

Each combat team of 44 (NLD) Mechanized Infantry Battalion has a tank platoon of formidable Leopard 2A6 tanks from Pantserbattalion 42 attached. Note the NATO three-color camouflage scheme.

The Netherlands and Germany are the only countries who have so far acquired this latest version of the Leopard. The platoon is of standard NATO size with four tanks.

One of the most interesting units is the 101st (NLD) RPV Battery, which is equipped with the French-manufactured Sagem RPV. The unit is organized around one command platoon and four launching system platoons. Each platoon has eight "birds" ready to take off.

The Sperwear drone has an autonomous flying time of four hours, and it can fly at an altitude of between 300-3000m at a speed of 90km/h. It takes four hours to set up the whole system.

The Spanish Ejercito provides NRF-4 with 24 Air Defense Battery, which is equipped with French-manufactured Mistral MANPAD missiles fitted on the rear platform of a Spanish-built HMMWV RAMTA. Here, a platoon of four vehicles is protecting a river crossing site.